New Zealand SHEEP and their wool

D. C. Dalton

David Bateman

Photographs by D. C. Dalton except on pages 2-3, 6, 11, 13 (bottom), 16-17, 24, 27 where they are by Geoff Moon, and the photograph on page 19 is courtesy of the New Zealand Wool Board.

First published in 1991 by
David Bateman Ltd
'Golden Heights', 32-34 View Road, Glenfield, Auckland,
New Zealand

Copyright © 1991 David Bateman Ltd

This book is copyright. Except for the purpose of fair review, no part may be stored or transmitted in any form or by any means, electronic or mechanical, including recording or storage in any information retrieval system, without permission in writing from the publisher. No reproduction may be made, whether by photocopying or any other means, unless a licence has been obtained from the publisher or its agent.

ISBN 1-86953-065-9

Typeset by Typocrafters Ltd
Printed by Colorcraft

Contents

History	4
Early sheepfarming in the North Island	6
Early sheepfarming in the South Island	7
Breeds	8
Farming in the North Island	10
Farming in the South Island	12
The sheepfarming year	14
Dogs	22
Transport and sales	25
Wool stores, auctions and wool classing	26
Carpet making	26
Craft — skins, spinning, weaving	28
Show days and Sheepworld	30

History

Millions of years ago New Zealand became an isolated island rich in birds, rather than other animals. All the farm animals you see in New Zealand today had to be introduced.

The Maori from Polynesia, who arrived around 1350 AD, introduced only a kind of dog and a type of rat, and it wasn't until European settlement in the early 1800s that sheep were successfully introduced.

Captain James Cook, the British explorer, earlier brought two sheep from Australia but they both died.

But the word spread that these islands in the South Pacific had great farming potential. So more sheep were brought here by European settlers in the early 1800s from the newly developing land of Australia.

There was one big problem for the early sheep farmers — the land was covered in dense bush which had to be cleared to grow pastures. So began sheep farming as you see it today in New Zealand.

Mustering in the North Island.

The first sheep to be farmed properly in New Zealand were fine-woolled Merinos from Australia. These sheep were native to Spain and thrived in the hot, dry conditions of Australia. But they were not suited to the warm, humid North Island of New Zealand, where European settlement rapidly developed. However, the potential for sheep was obvious as a means of feeding and clothing the early settlers and their families.

Soon the opportunity was also seen to export wool to markets on the other side of the world. The sheep population grew rapidly, and by the 1880s there were 13 million sheep in New Zealand. Sheep of many different types from many parts of the world were introduced. Many did not survive, the ones that really thrived were sheep from Britain, noted for their meat and wool.

Today, New Zealand has about 60 million sheep — about 20 sheep for every person in the country. We say with some pride that New Zealand's prosperity has come 'from the sheep's back' and meat and wool are still very important in our export economy. New Zealanders are very proud of the sheep and what it has allowed them to achieve.

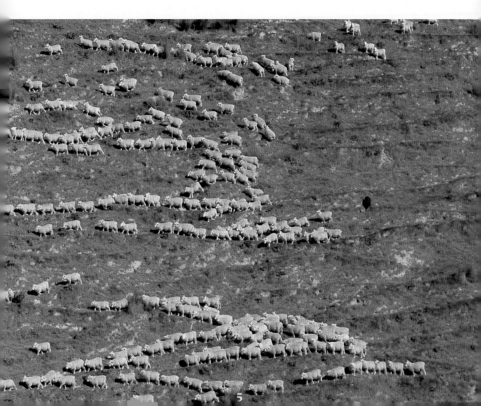

Early sheepfarming in the North Island

Wool was the main reason for the country's early prosperity. This was certainly the case in the North Island and there were some simple reasons for this. Wool did not deteriorate on the long sea voyage back to the markets of Europe, and could be easily packed into the holds of sailing ships.

Wool packed in large bales was easy to handle. In the early days of transport, before there was a good road system, wool bales had to be hauled from remote farms by hand, by teams of oxen and horses, and on boats along rivers.

On coastal farms with direct access to the sea, wool bales were loaded on to barges from horse-drawn carts which went right out into the surf. These barges then delivered the bales to the ships anchored offshore.

Another reason was that the harvesting of wool from the sheep was easy for the early farmers. All it needed was a pair of hand clippers or shears. The skill to use them came with the first shepherds from Europe.

Below *A pastoral scene from the King Country.*
Below right *Sheepfarming in Southland.*

Early sheepfarming in the South Island

In the South Island, sheep farming developed more quickly because there was not so much forest to be cleared as in the North Island.

The early farming pioneers left Christchurch and walked across the scrub-covered Canterbury Plains towards the hills and snow-capped mountains on the horizon.

They found some unique open high country, covered in tussock grass that sheltered other grasses and plants which sheep relished.

The high rainfall of the North Island contrasted with this drier South Island high country where long dry periods were not uncommon. There was also snow in the winter, sometimes right down to the plains. It was a totally different farming world and it was here that the Merino breed of sheep began to reign supreme.

The South Island quickly became a massive sheep station producing wool for the growing textile trade in Europe. However, in the 1880s wool prices began to drop. Just in time, the industry then got a boost from the development of refrigeration making it possible to export meat as well as wool. This heralded a new era for sheep farmers. It was a good example of how farmers used new technology to develop their industry, and this continues today. New Zealand is a world leader in the science and practice of sheep farming.

Romney rams   Coopworths

Breeds

Two things about sheep are very noticeable when travelling through New Zealand today. First, you will see different kinds of sheep, and second, you will notice that their wool varies in length. Some sheep look very clean and white when they have just been shorn.

New Zealand is the world's main producer and exporter of wool for carpets. This is referred to as 'coarse, strong or crossbred' wool. It is about one and a half times as thick as the 'fine' wool of the Merino.

This coarse wool comes from sheep breeds specially developed for New Zealand conditions. They are referred to as 'dual-purpose' (meat and wool) sheep. They produce an average of 5 kg (11 lb) of wool and their lambs at 100 days old weigh 27–30 kg (60–66 lb). These lambs then produce carcasses of about 13 kg for export.

The **Romney** can be seen all over the country, from the rich lowland pastures right up to the hills. It is the basis of all sheep farming in the North Island and in most of the lowlands of the South Island.

The **Coopworth** breed was developed by crossbreeding Romney with Border Leicester. The **Perendale** was developed from the Romney, crossbreeding with the Cheviot, for the steeper hill country.

The **Corriedale** was developed by crossing the English Leicester and Lincoln with the Merino, for the drier hills and plains of the South Island.

As the demand for specialist carpet fibre grew, farmers became interested in breeding sheep specially for this trade. The **Drys-**

The Drysdale breed   *Perendales*

dale is the best known breed of this type with its coarse fleece and large curly horns.

The **Merino** is the famous high-country sheep. Its fine and superfine wool is in demand for clothing. New Zealand Merinos produce some of the finest fibre in the world and this is highly sought after by Japanese textile mills.

Other sheep are particularly used to breed lambs for the meat trade. These are called 'Down' breeds and are used as rams (males) to cross with Romney ewes (females), usually after the Romney ewe is five years old and has finished her life on the hills. She is brought down to the lower pastures for one more crop of lambs. Some of the ram breeds used to mate with these ewes are the **Southdown**, the **Suffolk**, the **Polled Dorset**, the **Hampshire** and the **Wiltshire**. They were all developed from similar British breeds.

Home spinning and weaving, which was an essential part of the life of the early pioneers, has now developed into a very important craft in New Zealand. As a result, the odd black and coloured sheep born in farmers' flocks has been kept and bred for this special homecraft market. Many of these coloured sheep can now be seen in small flocks throughout New Zealand. Coloured sheep can arise from white-woolled parents in any breed because of minor genetic differences.

A climate which is ideal for pasture growth also is ideal for weeds. Some European plants which were introduced for hedges (e.g. gorse) have become major problems. Chemical weed control is used, but farmers also use the sheep themselves, or goats, to act as 'pasture weed-control machines'. This is done at times of the year when the stock do not have to be at peak production and can be run together in large mobs, which will eat out plants they would not normally graze.

Farming in the North Island

The main feature of the North Island's climate is a high rainfall, up to 1,500 mm (60 inches) a year. This has allowed sheep farming to develop to a very high intensity, as it produces rich healthy pasture for most of the year. Winter frosts are not common, and when they do occur the rest of the day is usually bright, warm and dry.

North Island pastures are mainly ryegrass and clover, which grow fairly evenly most of the year. This allows the farmers to keep high stock levels (an average of 12–15 ewes per hectare, or 5–7 per acre). Sheep in the North Island are never housed in winter and never fed grain, although they may be fed hay or silage in winter or during a drought.

An average-sized farm in the North Island runs about 1,500–2,000 breeding ewes, and also has 150 cattle. Such a farm is usually run by the farm owner, with help at busy times from the family — especially when lambing and shearing.

Most North Island sheep farms are on the rolling to steep hill country, because the valuable lowland is used for dairy farming and horticulture. This makes farming physically hard work for shepherds, dogs and horses. On some farms, horses have been replaced by the farm motorbike — especially with the younger shepherds.

On steep hill country, aeroplanes are used to spread fertilizers (lime and phosphate) to make grass grow well. When new pastures are sown, farmers use large numbers of sheep to press the seed into the ground with their sharp hooves. This technique was developed by the early pioneers who broke in the bush for farming.

Left *The North Island climate allows farmers to keep high stock levels.*
Above *Sheepfarming in the Waikato.*

Farming in the South Island

There are two main kinds of sheepfarming in the South Island. One is lowland farming, rather like that in the North Island. The other, more important kind, is the high-country sheep station, where some very large Merino and Corriedale flocks are farmed. These farms run 5–12,000 sheep on fairly poor land which can carry only 2–6 ewes per hectare (about 1–3 per acre). Rainfall is low (300–400 mm or 12–16 inches a year) and long dry summers are common.

On the Merino farms the sheep graze right up to the snowline of the high peaks in summer (up to 2,000 m or 6,000 feet) where they seek out the alpine herbs and grasses.

Mustering (gathering) these sheep for shearing and to bring them down for winter before the snow arrives is a major part of the farming routine. Most of the shepherd's life used to be spent on horseback on the easier slopes, and on foot on the steeper country. However, today the modern high-country farmer uses a light aeroplane or helicopter to get quicker access for shepherds and their dogs to where the Merinos graze. Farms also still have shepherds' huts in the mountains, which are used as bases for mustering.

Even when these high-country sheep reach the safety of the farm's fenced pastures, they may still be subject to cold winters and deep snow. To help them through the winter they are given supplements of grain, hay and silage. Sheep are never housed; they spend all their life outdoors. Shepherds have to be on constant watch for snowfalls which could bury their sheep.

The worst problem for farmers and shepherds on these high-country farms is late snowstorms when lambs are newly born; or just after shearing, when the sheep feel the cold. There have been some terrible losses of sheep at these times, and they can devastate the farm as it takes many years for the flock to rebuild its numbers.

Above right *A sheep farm in the Otago hill country.*

Right *Lush pasture in Nelson.*

The sheepfarming year

Autumn

The sheep farming year really starts in autumn. In New Zealand this is usually late February in the North Island and late March to April in the South Island.

This is the time when farmers bring in their ewes which they have decided to keep for the coming season, and prepare them for mating with the rams. Next year's lamb crop depends a great deal on how heavy the ewes are at mating, and how well the farmer cares for the flock over the mating period. Sheep become ready for breeding as the days become shorter. The reduced length of autumn days triggers the ewes' breeding cycle and has similar effects on the rams. Thus nature ensures that, when the ewes give birth 5 months later, in the spring, there will be plenty of fresh pasture to provide a good milk supply for the hungry, growing lambs.

Farmers usually keep 50–100 ewes with one ram for two periods of 17 days, which is the length of the ewe's breeding cycle. To tell which ewes have been mated, the ram is fitted with a mating harness which holds a coloured crayon. After the ram has mated, this leaves a coloured mark on the rump of the ewe. The farmer then looks after the ewes during the winter and early spring, according to their stage of pregnancy. The ewes which are mated first and will lamb first, need more careful management.

Winter

Winter is the quietest time on a sheep farm. General farm maintenance is carried out; fences are repaired, drains are cleaned out and swamps are drained so lambs will not get drowned or lost in spring. Water troughs and the farm water supply are also looked after. Repairs will be carried out to the sheep yards and woolshed so that everything is ready for shearing in the summer.

Gates that do not open easily, and holes in yards where lambs can escape, can be very frustrating for hard-working shepherds in the middle of shearing or other farm jobs. Winter is the time to fix all these things.

A sheep flock in typical North Island hill country.
Overleaf *A pleasant early summer's day in Otago.*

Feeding the ewes well during winter can be a problem because grass and clover growth slows down. It is important to ensure a good supply of feed is ready for the ewes and their new-born lambs in the early spring. So farmers restrict the amount of grazing offered to the flock in winter and provide supplementary feed like hay and silage. This is done by running all the ewes together and slowly moving them around the farm. This is called 'rotational grazing.' Making the ewes go a little hungry now means they will have more feed when they most need it — after lambing in the spring.

During the 5 months of pregnancy during the winter, the ewes' nutritional needs are at their lowest. Farmers plan to maintain the ewe's body weight during this time.

Left *Ewes may have twin or triplet lambs...*
Above *...and sometimes another source of milk is needed.*

Spring

This is lambing time and probably the busiest time of the year. In recent years New Zealand farmers have developed breeds of sheep that give birth to their lambs and look after them without needing assistance. Cows also calve in spring so there is little spare time for people on the farm.

In some flocks, the ewes have many twin and triplet lambs, so shepherding has to be more vigilant and the ewes need extra feed to provide the extra milk for 2 or 3 young lambs. At present the average lamb production is one per ewe per year but farmers are always striving to increase this.

About 6–8 weeks after birth the lambs have to be docked. This is where the lambs' tails have to be removed so that they will remain clean when they start eating lush pasture. It is also the time when surplus male lambs which are not needed for breeding are castrated.

Above *Wool bales ready for transporting represent many hours of hard work.*

Right *Sorting wool is a job for experts.*

Summer

Shearing is the next major job on the farm, usually taking place once a year in early summer. Both the ewes and the lambs are shorn. However, some farmers may shear twice a year or 3 times in 2 years.

Shearing and the associated handling of the wool is usually done by a group of experts called a 'gang' who work on contract. The farmer musters the sheep and all the rest is done by the contractor.

An average New Zealand shearer is expected to shear between 300–400 sheep in an 8-hour day. An average fleece will weigh 4.5 kg (10 lb). Some shearers work a standard 9-hour day and the current world record is nearly 600 sheep in 9 hours. This is an incredible achievement of skill and fitness. The finished job is a stack of wool bales in the woolshed waiting for transport to the woolstore.

Handling of the wool after shearing needs great care, as today this wool may not be touched again until it goes into the textile machine on the other side of the world. This means that correct sorting and the removal of dirty or unsound wool has to be done on the spot by experts who know what the market demands.

New Zealand is noted worldwide for the development of better shearing techniques. The standards set by modern shearers are extremely high, and speed is not allowed at the expense of quality of work and care for the sheep.

About half of New Zealand's wool production (known as 'the clip') is scoured (washed) before it is baled and shipped to market. This reduces the work the processor has to do to the wool and reduces the shipping weight with the removal of the natural grease or lanolin.

Summer is also the time to dip the sheep, which protects them from parasites on their skin. To do this, the sheep are put through a shower or swim bath.

End of Summer

At the end of summer the lambs (now about 4 months old) are weaned from the ewes.

This is also a time for the shepherds to give the lambs a dose of drugs (called a 'drench') for internal parasites before putting them on to clean grazing land specially saved up for them. It is important that they settle and grow quickly.

Now is the time for removing or 'culling' any ewes which did not wean a lamb, have lost teeth, or have some health problem that will prevent them producing a good lamb the next season. Any ewes over 5 years of age are also culled and usually sold to other farmers on better land for one more lamb crop.

These 5-year-old ewes will be replaced by the same number of young, 18-month-old ewes which the farmer has bred as replacements. This allows a constant number of sheep to be kept in the flock. The older ewes are always more productive than the young ewes. Any surplus stock are sold. There are often special sales for stock of different ages.

This is also the time when a sheep farmer checks the rams to see they are all in good health before the autumn. Any replacement rams are usually purchased from a breeder who has stock specially selected for high performance.

Right *Sheepfarming in Hawkes Bay.*

Dogs

Without the working sheepdog there would be no New Zealand sheep industry. This is so well appreciated that a monument has been erected near Lake Tekapo in gratitude to these great animals.

When the first farmers and shepherds came to New Zealand they brought with them Border Collie dogs from Britain. These were then bred to suit local conditions so that now New Zealand has some very special dogs.

There are 2 main types. First is the black, white and tan 'heading dog' which is sent around mobs of sheep to 'head them off' or stop them escaping. Some of these are called 'eye' dogs because of their ability to stare fixedly at the sheep, which stops them from running away. New Zealand heading dogs have longer legs and have a shorter coat than the British Border Collie of today.

The other type of working dog which is unique to New Zealand is called the 'huntaway' because it is used to drive or 'hunt' sheep away from the shepherd. It will also head sheep, but not quite the same way as the true heading dog. Huntaways are big, robust, black-and-tan-coloured, loudly barking dogs. The barking is needed to drive sheep out of gullies in steep hill country and move big mobs at a great distance from the shepherd.

Huntaways are also used to work in sheep yards and woolsheds and will force sheep into pens by jumping on their backs. Dogs that will do this are called 'backing dogs.' So a good huntaway will head, hunt, bark and back.

Shepherds show off the ability of their dogs at dog trials, which are held all over New Zealand. They are great occasions for the shepherds to meet and talk, as well as test their dogs. There is a special course for heading dogs where they have to gather three sheep and help the shepherd put them in a pen. This exercise is a difficult challenge for the dog and shepherd alike. Then there are huntaway trials, where they have to drive three sheep up a very steep hill through markers while the handler stands at the bottom. The course may be straight or follow a zig-zag.

Above left *A good heading dog is a vital part of any shepherding team.*
Above right *Robust and rowdy — the New Zealand 'huntaway'.*

Below *Working sheepdogs in action.*

A heading dog shows its skills at the local dog trials.

Dogs which gain high marks in local and regional trials then compete in the national trials. The top dogs are then in demand for breeding, and this ensures their superior qualities are passed on to the other working dogs in the country.

Heading dogs and huntaways are rarely crossed to combine their qualities. Shepherds prefer to keep the breeds separate and work them as a team.

A shepherd on a typical hill-country sheep and cattle farm would have a team of 5–7 dogs. A 5-dog team is usually made up of 2 heading dogs and 3 huntaways. The shepherd may also have a puppy of each breed to be trained. As the dogs get to about 6–7 years old, they are usually sold or given to farmers with small blocks of land down on the lowland — a retirement job for the dog that has made such a great contribution on the hills and wide open spaces.

Transport and sales

New Zealand is made up of two long thin islands, with many smaller islands that are farmed with sheep and cattle. Long-distance transport of stock therefore is a very important part of our farming industry.

In the early days, before highways and high traffic density, stock were made to walk to the sales. This was called 'droving.' Today's heavy traffic has almost banished stock from the roads, and even where farms are divided by a highway, underpasses have had to be built for stock to cross.

Sheep are transported in large double-decked trucks and trailers. Inside the truck the sheep are penned in small groups to ensure their safety during travel. They are unloaded on to special ramps so they can walk out safely without jumping and injuring themselves.

Truck drivers are also very experienced at handling stock. A driver usually has a huntaway dog to help load and unload the sheep. The dog travels in a special box on the side of the truck and seems to enjoy the journey between spells of work. Dogs regularly travel with their head out of the box, looking at passing traffic and barking.

The saleyards provide the focal point for the local sheepfarming community.

New Zealand sheep sales are held at special venues with permanent yards and facilities. Stock are taken to these saleyards where they are penned. Each pen has a number and the auctioneer and buyers move along to each pen in turn. The auctioneer and any assistant stock agents may stand among the sheep or on a catwalk above the pens. Selling is a lively affair, with the auctioneer trying to get that extra bid for his vendor. The sale is a great social occasion too, especially for farmers who live in the back country on remote farms.

Wool stores, auctions and wool classing

Wool is a very special fibre and has some unique qualities. Everyone in New Zealand appreciates this, and it can be seen in the way wool is shorn, sorted, stored and transported.

Sheep must be dry when they are shorn because wool absorbs moisture — a great asset to the wearer of a woollen jersey! If the sheep are damp before shearing the farmer will put them in a paddock on a ridge where they can get plenty of breeze to dry them off.

The fleeces are packed about 50 to a bale, which weighs about 160–180 kg (350–400 lb) and is branded with the farm identification and the type of wool inside. This is very important. At the wool store each bale is sampled and tested. The wool is then sold with a certificate giving detailed specifications.

Buyers come along and inspect these samples, and then decide what to buy when the auction takes place a few days later. Buyers now do not need to go to the long stacks of bales and pull the wool out for inspection the way they used to do.

Carpet making

New Zealand carpets are valued worldwide for the qualities of the wool that goes into them. A vitally important part of a wool carpet, especially if you want it to last a long time, is its ability to stand up to regular wear. This comes from the resilience in the wool fibre — a quality some New Zealand sheep have above all others. You can see this if you squeeze a handful of wool tightly in your hand and then release it. Does it bounce back to its original shape? Good carpet wool should.

This is where the special carpet-wool breed, the Drysdale, comes into its own. Other breeds like the Romney, Cheviot and Perendale have fibres that provide bulk to the carpet mixture — this exact mix being a closely guarded secret of carpet manufacturers.

Carpet wool has to be suitable for modern high-speed machinery; the days of making carpets on cottage looms are long gone. This means the fibres have to be strong with no weaknesses in them — a quality which comes from good feeding and management of the sheep.

To accept a wide range of dyes, especially pastel shades, carpet wool has to be a good white colour and not stained by wet weather and poor management on the farm. This explains why shepherds have to be aware of what manufacturers and buyers in other parts of the world need from their sheep.

Sheepfarming in the hill country north of Auckland.

Craft — skins, spinning, weaving

New Zealanders have shown great skill in the art and craft associated with wool. Making warm woollen clothes for everyone in the family from the wool of their sheep was part of the daily life of the early settlers.

There are now many local and national woolcraft festivals in New Zealand where the best spinners, weavers and artists demonstrate the high international standards they have achieved.

People often start by spinning the wool from a black sheep or a black lamb given to them and reared as a pet. This is then knitted into a work jersey to be worn on the farm. The craft often develops from this simple start. Mothers teach children and the craft is encouraged at most country schools, where local craft people offer their support.

All over New Zealand there are spinning groups. Spinning wheels and small looms can be readily purchased to suit all needs. There are companies who specialise in larger home looms for craftspeople who progress to weaving material which may then be made into fashion garments. Many of these garments appear in national fashion shows.

Sheepskin craft has also developed — again a skill which was common in pioneering days when farmers cured their own sheep skins. Some home curing is done but people mainly prefer to buy professionally-cured skins. These are usually made from the pelts of lambs.

Below left *Black and coloured sheep are sought after by spinners, weavers and artists.*
Below *Wool is the perfect medium for the exciting textures and brilliant colours of this jersey.*

Show days and Sheepworld

Farmers and rural people like to get together in New Zealand. Many of them live in remote areas and they enjoy the many opportunities families have to meet.

One of the most popular is the Pet Day at their local schools. This is where children bring a pet lamb to school on a special day where it is judged by a local farmer for such qualities as rearing, leading, breed type and its ability to come when called. Ribbons are presented for the best lambs. The winners at each school go on to district finals held at the local agricultural shows.

The agricultural show is held in summer and is a highlight of the farming year. They are run by Agricultural and Pastoral (A&P) societies and are a great opportunity for country and city people to meet. This is now very important as only about 6% of the New Zealand population is employed in farming, but many more are involved in industries supporting agriculture.

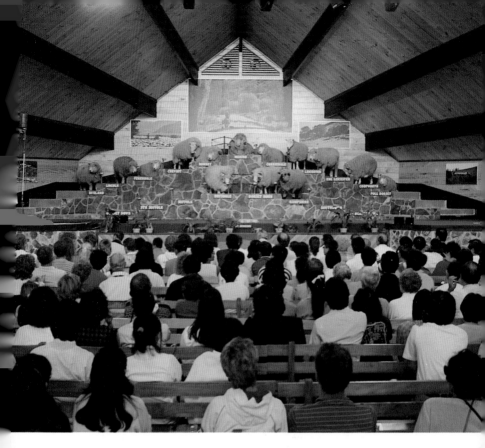

Above left *Pet lambs on show at a School Pet Day.*
Above *The Agrodome in Rotorua provides a glimpse of New Zealand sheepfarming.*

At the local shows, farmers can see the livestock which have been judged as the best example of their breed, and they can see all the latest farming equipment. Often one of the early events in the dog trial calendar is held at the local A&P show. The dog trials then run into the autumn and early winter.

There are also special machinery and equipment field days to show farmers new developments. They are held in the winter when farmers have more time available to spend off the farm.

Sheepworld at Rotorua is a very special feature. The story of sheep — the breeds, their history, shearing and working a huntaway dog — is told in a unique way. This way of showing the importance of sheep was developed in New Zealand and is now being copied in other countries of the world.

Above left *Shepherd recording the identity of lambs.*
Above right *A shepherd and his huntaway.*
Below *Sheepfarming in the North Island hill country.*